HOT AND COLD ANIMALS

SPOTTED EAGLE-OWL —OR— SNOWY OWL

BY ERIC GERON

Children's Press
An imprint of Scholastic Inc.

A special thank you to the team at the Cincinnati Zoo & Botanical Garden for their expert consultation.

Library of Congress Cataloging-in-Publication Data
Names: Geron, Eric, author.
Title: Hot and cold animals. Spotted eagle-owl or Snowy owl / by Eric Geron.
Other titles: Spotted eagle-owl or Snowy owl
Description: First edition. | New York : Children's Press, an imprint of Scholastic Inc., 2022. | Series: Hot and cold animals | Includes index. | Audience: Ages 5–7. | Audience: Grades K–1. | Summary: "NEW series. Nonfiction, full-color photos and short blocks of text to entertain and explain and how some animals with the same name can survive in very different environments"—Provided by publisher.
Identifiers: LCCN 2021044787 (print) | LCCN 2021044788 (ebook) | ISBN 9781338799422 (library binding) | ISBN 9781338799439 (paperback) | ISBN 9781338799446 (ebk)
Subjects: LCSH: Bubo—Juvenile literature. | Snowy owl—Juvenile literature. | Owls—Juvenile literature. | Habitat (Ecology)—Juvenile literature. | BISAC: JUVENILE NONFICTION / Animals / Birds | JUVENILE NONFICTION / Animals / General
Classification: LCC QL696.S83 G47 2022 (print) | LCC QL696.S83 (ebook) | DDC 598.9/7—dc23
LC record available at https://lccn.loc.gov/2021044787
LC ebook record available at https://lccn.loc.gov/2021044788

10 9 8 7 6 5 4 3 2 1   22 23 24 25 26

Printed in the U.S.A. 113
First edition, 2022

Book design by Kay Petronio

Photos ©: cover right: Benjamin Olson/Minden Pictures; back cover right: Guy Edwardes/Minden Pictures; 1 right: Benjamin Olson/Minden Pictures; 2 right: Guy Edwardes/Minden Pictures; 8–9: Jim Cumming/Getty Images; 10 top right: Guy Edwardes/Minden Pictures; 12–13: Herbert Kratky/imageBROKER/Alamy Images; 14–15: by Mark Spowart/Getty Images; 16 top right: Guy Edwardes/Minden Pictures; 16 center: Kierran Allen/Dreamstime; 16 bottom: Jim McMahon/Mapman ©; 17 left: Jim McMahon/Mapman ©; 18–19: Peter Steyn/ardea.com/age fotostock; 20–21: Michio Hoshino/Minden Pictures; 22 top right: Guy Edwardes/Minden Pictures; 22 main: Jean-Paul Chatagnon/Biosphoto; 23: tatianaput/Getty Images; 26–27: Winfried Wisniewski/Minden Pictures; 28 top right: Guy Edwardes/Minden Pictures; 29: Jspix/imageBROKER/Biosphoto; 30 right: Benjamin Olson/Minden Pictures. All other photos © Shutterstock.

SPOTTED EAGLE-OWL

SNOWY OWL

CONTENTS

MEET THE OWLS

↗ SPOTTED EAGLE-OWL

Spotted eagle-owls and snowy owls are very different kinds of owls. Spotted eagle-owls live in Southern Africa, where it's warm. They like to hunt at night and take baths in the rain.

The spotted eagle-owl is also known as the African eagle-owl.

FACT

Snowy owls live in the chilly **tundra**. They like to hunt during the day and keep dry.

SNOWY OWL

FACT

The snowy owl is also known as the great white owl.

NIGHT VISION

Since the spotted eagle-owl hunts at night, it has large eyes that let it see in the dark.

EAR FEATHERS

A spotted eagle-owl has fluff over its ears called ear tufts.

RAZOR BEAK

Without teeth, this powerful beak can grip food and tear it into smaller pieces if it's too large to swallow whole!

WHITE PATCH

Spotted eagle-owls have a special white patch on their throats to help other spotted eagle-owls see them in the dark.

SILENT FLIGHT

Having soft wing feathers means the spotted eagle-owl can fly silently while hunting.

FACT The round face shape of owls helps them to hear better.

SHARP CLAWS

Strong feet and talons help the spotted eagle-owl catch its prey.

A spotted eagle-owl can weigh 1 to 2 pounds (0.5 to 1 kg).

It has a brown body and dark stripes on its tail and wings.

7

SNOWY OWL CLOSE-UP

A snowy owl can weigh 3–½ to 6–½ pounds (1.6 to 3 kg).

Males have all-white feathers, while females have little dark markings on their white feathers.

TAKE WING

These strong wings help the snowy owl catch fast prey.

FUZZY SLIPPERS

A snowy owl's feathers are thick and fluffy to keep them warm in the cold and silent while flying.

SIGHT AND SOUND

Excellent eyesight and hearing help the snowy owl easily locate prey.

BEAK BRISTLES

The special bristles on a snowy owl's beak can sense when objects are nearby.

OPEN WIDE!

Since it doesn't have any teeth, the snowy owl tears its food into smaller pieces or swallows it whole!

FACT Snowy owls usually stay close to the ground when they fly.

Spotted eagle-owls are medium-sized compared to other owls. Snowy owls are one of the biggest types of owls. The spotted eagle-owl has ear tufts

SPOTTED EAGLE-OWL

while the snowy owl does not. The snowy owl weighs almost double the weight of the spotted eagle-owl and is nearly twice as tall. Their **wingspans** are similar in length, at around 3 to 5 feet long (1 to 2 m).

SNOWY OWL

FACT

Spotted eagle-owls and snowy owls both have yellow eyes.

In the summer, the weather in Southern Africa can reach temperatures of 82°F (28°C)!

HOME, HEAT HOME

Southern Africa is a hot and **tropical** place. Spotted eagle-owls can survive in many types of **habitats** there such as deserts, woodlands, and **savannas**. Spotted eagle-owls take baths often and sometimes even stand out in the rain during thunderstorms. During the day, they live in places like trees, caves, and under bushes.

LET IT SNOW!

Snowy owls mostly live on the icy tundra of the Arctic. There are no trees there. It's mostly snow.

They like to stay on the ground at night and fly around during the day. They try not to use up too much energy in order to stay warm in their cold and dry environment.

FACT

The temperature in the Arctic can dip as low as −40°F (−40°C) in the winter!

WHO-WHO'S THERE?

Spotted eagle-owls and snowy owls live in very different places. The tropics of Southern Africa are sunny and warm. The Arctic is freezing cold. Spotted eagle-owls mostly live close to humans.

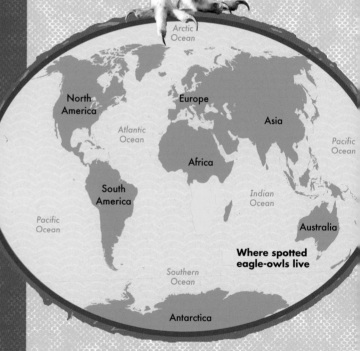

Arctic
Ocean

North
America

Europe

Asia

Atlantic
Ocean

Pacific
Ocean

Africa

South
America

Indian
Ocean

Pacific
Ocean

Australia

Where spotted eagle-owls live

Southern
Ocean

Antarctica

SPOTTED
EAGLE-OWLS

Snowy owls mostly live far away from humans. The spotted eagle-owl is brown like a tree branch. The snowy owl is white like fresh snow. Their different bodies help them **camouflage**, or blend, into their surroundings.

SNOWY OWLS

Arctic Ocean

North America

Europe

Asia

Atlantic Ocean

Pacific Ocean

Africa

South America

Indian Ocean

Pacific Ocean

Australia

Where snowy owls live

Southern Ocean

Antarctica

FACT

A group of owls is called a parliament.

17

Spotted eagle-owls and snowy owls are **carnivores**, which means they eat meat.

FACT

NIGHT OWL

Although it is bright and sunny where they live, spotted eagle-owls rest during the day and spend more time on the move at night! That's because they are **nocturnal**. Nocturnal describes an animal that is active during the night.

In the darkness, the spotted eagle-owls hunt for food. They mostly eat insects, rodents, and birds.

The spotted eagle-owl swoops down from the night sky and scoops up its prey in its claws.

PECKING ORDER

Snowy owls are **diurnal**. This means they hunt during the day. Snowy owls mostly eat **lemmings**. They also eat rabbits, birds, and fish. When food runs low, the snowy owl travels to places with more food.

FACT A snowy owl can eat up to 1,600 rodents in a single year!

SPOTTED EAGLE-OWL

INSECT

Spotted eagle-owls eat a lot of different things. Snowy owls only eat a few things. This may be because the woodlands, grasslands, and savannas have plenty of different

Eagles, dogs, cats, and humans are some of the spotted eagle-owl's **predators**.

FACT

food options available. Spotted eagle-owls stay in one place, while snowy owls **migrate** and make their homes wherever there is food. After swallowing food whole, owls cough up parts they can't digest, like the bones and fur. The spat-up food is called a **pellet**. When enemies are near, each owl lets out a hoot to defend itself.

MOUSE

SNOWY OWL

FACT The predators of the snowy owl include foxes, wolves, and sometimes humans.

Spotted eagle-owlets hatch with gray eyes that turn yellow after two weeks.

FLUFFY OWLETS

Spotted eagle-owl adults find a place on the ground to build their nest. They often use this same nesting place year after year. Spotted eagle-owl mothers lay two to three eggs per **clutch**. A clutch is the total amount of eggs laid at a single time. The eggs hatch after 30 days. Baby owls are called **owlets**.

25

WISE OWLETS

Snowy owl adults also build their nests on the ground. They usually lay three to 11 eggs per clutch. The eggs hatch after 32 days. Snowy owlets have soft white **down feathers** at first, which turn gray. They will leave the nest on their own after one month.

Snowy owl adults build their nests in different spots year after year.

SPOTTED EAGLE-OWLET

AWW... OWLETS!

Even though they grow up to be different adult owls, spotted eagle-owlets and snowy owlets are the same in many ways.

Mother spotted eagle-owls and mother snowy owls each sit on their eggs to keep them safe and warm until they hatch.

They are both born in nests on the ground and learn to fly when they are one and a half months old. They stay with their parents for about 10 to 12 weeks after they learn how to fly.

SNOWY OWLETS

YOU DECIDE!

If you had to choose, would you rather be a spotted eagle-owl or a snowy owl? If you like tropical weather and don't mind moving around at night, maybe you would choose to be a spotted eagle-owl. If you like cold weather and traveling over the snow during the day, you may prefer being a snowy owl!

FACT There are 268 different types of owls.

GLOSSARY

camouflage (KAM-uh-flahzh) – to disguise something so that it blends in with its surroundings

carnivore (KAHR-nuh-vor) – an animal that eats meat

clutch (kluhch) – a group of eggs

diurnal (dye-YUR-nuhl) – active during the day

down feathers (doun FETH-urs) – the soft feathers of a bird

ear tufts (EER tuhfts) – skin covered in fluffy feathers that look like ears

habitat (HAB-i-tat) – the place where an animal or a plant is usually found

lemming (LEM-ing) – a small rodent found in the Arctic tundra

migrate (MYE-grate) – to move from one area to another at a certain time of year

nocturnal (nahk-TUR-nuhl) – active at night

owlet (oul-ET) – a baby owl

pellet (PEL-it) – a small, hard ball of something, such as food

predator (PRED-uh-tur) – an animal that lives by hunting other animals for food

prey (pray) – an animal that is hunted by another animal for food

savanna (suh-VAN-uh) – a flat, grassy plain with few or no trees

tropical (TRAH-pi-kuhl) – of or having to do with the hot, rainy area of the tropics

tundra (TUHN-druh) – a very cold area where there are no trees and the ground is always frozen

wingspan (WING-span) – the distance from one wing tip to the other wing tip

INDEX

ABOUT THE AUTHOR

Eric Geron is the author of many books. He lives in Los Angeles, California, with his tiny dog. If he had to choose, he would be a snowy owl.